The Mathematician's Thought Catalogue

Copyright © 2016
All rights reserved.

ISBN-13: 978-1537442754
ISBN-10: 1537442759

www.ingramcontent.com/pod-product-compliance
Lightning Source LLC
Chambersburg PA
CBHW071811200526
45169CB00017B/110